Using Resources

Lesson 1

How Do People Use Soil and Water Resources? 2

Lesson 2

How Can People Conserve Resources? 12

Harcourt
SCHOOL PUBLISHERS

Orlando   Austin   New York   San Diego   Toronto   London

Visit The Learning Site!
www.harcourtschool.com

Lesson 1

How Do People Use Soil and Water Resources?

VOCABULARY
renewable resource
nonrenewable resource
pollution

A **renewable resource** is a resource that can be replaced within a human lifetime. A renewable resource is reusable. Air and water are renewable resources.

A **nonrenewable resource** is a resource that cannot be replaced within a human lifetime. Soil is a nonrenewable resource. It can take thousands of years for soil to form.

Pollution is any change to the environment that can harm living things. Water pollution is one kind of pollution.

READING FOCUS SKILL
MAIN IDEA AND DETAILS

The **main idea** is what the text is mostly about. **Details** are pieces of information about the **main idea**.

As you read, look for **details** about how people use natural resources.

Natural Resources

Natural resources are materials found in nature that people use. Air, water, and soil are all natural resources. Many things are made from natural resources, such as books, CDS, and bikes. Gasoline, fuel oil, and coal are also natural resources. Natural resources make life on Earth possible.

How many natural resources can you see in this picture?

Some natural resources are reusable. Air and water are reusable. Natural resources are reusable and called renewable resources. A **renewable resource** can be replaced within a human lifetime. A renewable resource can be used again and again.

Some resources can be used only once. They are called nonrenewable resources. A **nonrenewable resource** cannot be replaced within a human lifetime. Soil is a nonrenewable resource. It can take soil thousands of years to form. Coal and oil take millions of years to form. Minerals as well as metals are also nonrenewable.

Focus Skill **Tell the difference between a renewable resource and a nonrenewable resource.**

How many renewable resources can you see in this picture? How many nonrenewable resources?

Air Pollution

Pollution is any change to the environment that can harm living things. *Air pollution* is one kind of pollution. Most air pollution is caused by burning fuels, such as coal and oil. The exhaust from cars, buses, and trucks is pollution. The fuels they use come from oil. When the fuels burn, they send harmful gasses into the air.

Smog is a kind of air pollution. Smog is a mixture of smoke, water vapor, and chemicals. It looks like a yellow or brown haze in the air. Smog usually comes from factories and cars and trucks.

▼ Air pollution in the form of smog can cause serious health problems.

Acid rain is another kind of pollution. Chemicals mix with water in the air. Then they form acids. The acid falls to Earth in rain. Acid rain can kill fish and damage trees and other plants.

Tell how the air can become polluted.

Pollution from vehicles ▼

7

Water Pollution

When harmful materials get into the water cycle, they cause *water pollution*. Some of these materials come from factories and mines. The factories and mines dump wastes into rivers, lakes, and streams. These wastes can get into the groundwater. Fertilizers and pesticides can pollute groundwater, too.

Wastes can spill or leak.

The wastes seep into underground water.

If the polluted water joins larger bodies of water, it pollutes them, too.

Sewage can also pollute water. *Sewage* is human waste. It is usually flushed away with water. If sewage gets into the water supply, it can make people sick.

There are laws to control water pollution. People must clean the water they use before returning it to the environment.

Focus Skill **Tell how water can become polluted.**

◄ Underground water supplies can be polluted by materials that soak into the ground.

The polluted water reaches the ocean and pollutes it.

9

Land Pollution and Misuse

Garbage in dumps and landfills can cause *land pollution*. Some garbage may contain harmful materials that get into the ground. Some things, like plastics, that are thrown away take a long time to break down.

Poor farming methods can cause land misuse. Unplanted land must be protected. If it is not protected, the soil may be carried away by wind and water. Soil is a nonrenewable resource. It can take thousands of years to replace.

Wastes from factories can also harm land. Some poisonous wastes are buried in large containers or drums. If the containers leak, the poisons get into the ground.

Unprotected soil can be carried away by rain and wind. ▼

People are working to stop land pollution. Most farmers use modern methods to protect the land. Communities and factories can get rid of wastes in ways that don't pollute the land. The government has also spent millions of dollars cleaning up polluted land.

Tell how the land can become polluted.

◀ Dumps and landfills can pollute the land.

Review

Complete this main idea statement.

1. Natural resources are materials found in _____ that people use.

Complete these detail statements.

2. Most air pollution is caused by burning _____.

3. Sewage can pollute the _____.

4. Unplanted land can be carried away by _____ and _____.

Lesson 2

VOCABULARY
conservation

How Can People Conserve Resources?

Conservation is the protection of natural resources. When you reuse or recycle something, you are conserving.

READING FOCUS SKILL
CAUSE AND EFFECT

A **cause** is what makes something happen. An **effect** is what happens.

As you read, look for the **effect** conservation has on reducing pollution.

Reduce, Reuse, Recycle

When you help to save natural resources, you are practicing conservation. **Conservation** is the protection of natural resources. One good way to conserve natural resources is to practice the three Rs—reduce, reuse, recycle.

When you *reduce*, you cut down on the resources you use. Electrical appliances, such as hair driers and air conditioners, use energy. When you use them less often, the need for energy resources goes down. The need to make less energy reduces pollution too.

Reduce your shower time.

Cut 5 minutes off your shower and save 20 gallons.

When you *reuse*, you use things that might have been thrown away. Reusing plastic food containers, means fewer resources are needed to make new ones. Reusing also means using items in new ways. You can wash out milk cartons and juice bottles. Then use them as planters or bird feeders. Reusing saves resources, reduces pollution, and saves space in landfills.

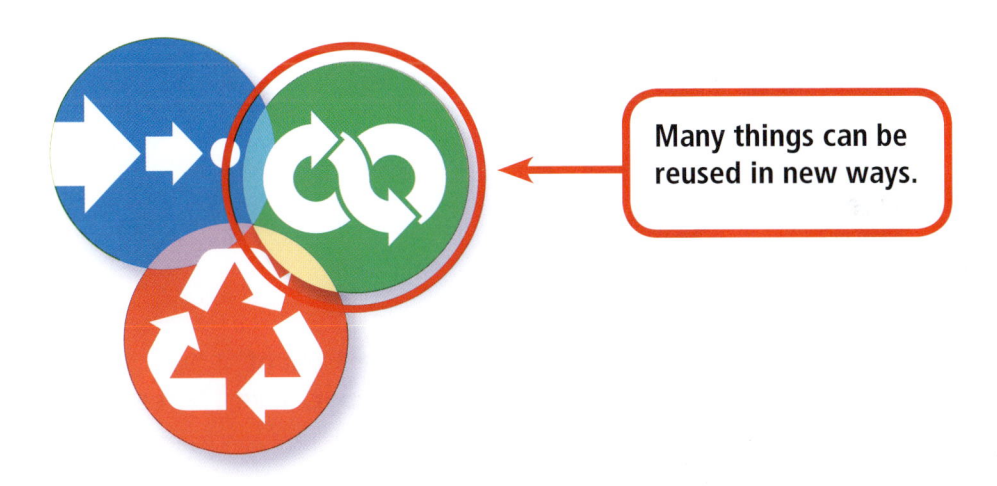

Many things can be reused in new ways.

When you *recycle*, items are changed into a form that can be used again. Aluminium, glass, and paper can be recycled. They can be ground up or melted down to make new products. When people recycle, energy is saved, too. Recycling saves resources and reduces pollution.

🌟 **Focus Skill** **Tell what effect practicing the three Rs has on conservation.**

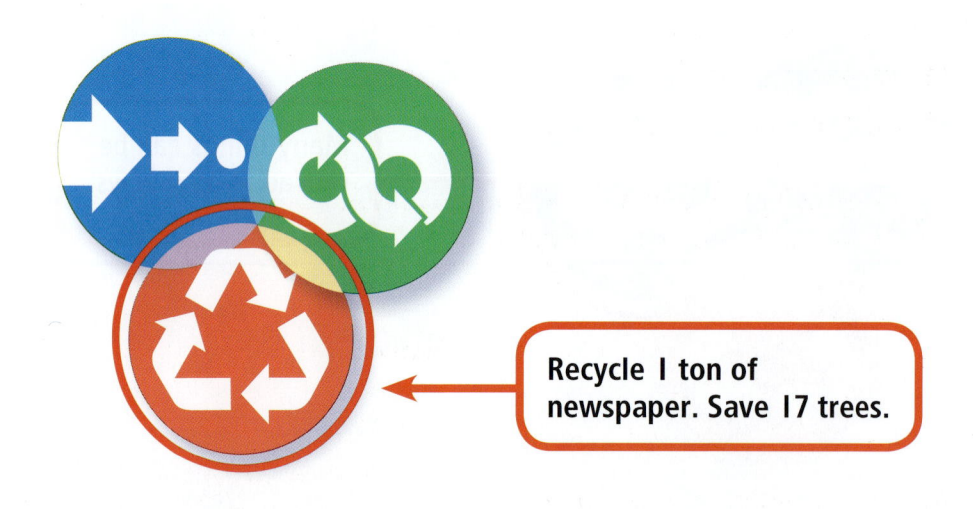

Recycle 1 ton of newspaper. Save 17 trees.

Water Conservation

Fresh water is a valuable resource. As our population grows, we use more and more water. Droughts, or times of little or no rain, affect some parts of the country.

As a result, everyone needs to conserve water. Some farmers have started growing crops that don't need much water. Others have stopped spraying crops with water. Instead, they use drip irrigation.

People can conserve water by taking shorter showers. They can remember to turn faucets off. They can use only the amount of water that they really need.

Focus Skill: Tell what effect growing populations and droughts have on water resources.

Native grasses, flowering plants, and shrubs use less water than lawns. ▼

Soil Conservation

Many farmers use methods that protect the soil. Some farmers use contour plowing to keep soil in place. Contour plowing is plowing around the slopes of hills. It takes the place of plowing up and down hills.

Some farmers use windbreaks to keep soil from blowing away. These are rows of trees or fences that block the wind.

Crop rotation is another way farmers help the soil. Crop rotation is changing crops that are planted from year to year. By rotating crops, farmers can rebuild the soil's nutrients naturally.

◀ Crop rotation can help control pests.

Contour plowing helps keep soil in place.

Intercropping also helps soil. Intercropping reduces the need for pesticides. When farmers intercrop, they plant certain crops near each other. This keeps some harmful insects from spreading.

Tell what effects farmers have had on soil conservation.

This is one kind of intercropping. A tall crop is planted next to a low crop that grows well in shade.

Review

Complete these cause–and–effect statements.

1. Using the three Rs, helps _____ natural resources.

2. One effect of reusing is the saving of _____.

3. When you _____, you cut down on the natural resources you use.

4. The effect of a growing population is that we use _____ and _____ water.

GLOSSARY

conservation (kahn•ser•VAY•shuhn) the preservation or protection of natural resources. the use of less of a resource to make the supply last longer

nonrenewable resource (nahn•rih•NOO•uh•buhl REE•sawrs) a resource that, once used, cannot be replaced within a human lifetime.

pollution (puh•LOO•shuhn) any change to the natural environment that can harm living things.

renewable resource (rih•NOO•uh•buhl REE•sawrs) a resource that can be replaced within a human lifetime; a resource that is reusable.